ANIMALS GROW

Lambs

by Anne Wendorff

BELLWETHER MEDIA • MINNEAPOLIS, MN

Note to Librarians, Teachers, and Parents:

Blastoff! Readers are carefully developed by literacy experts and combine standards-based content with developmentally appropriate text.

Level 1 provides the most support through repetition of high-frequency words, light text, predictable sentence patterns, and strong visual support.

Level 2 offers early readers a bit more challenge through varied simple sentences, increased text load, and less repetition of high-frequency words.

Level 3 advances early-fluent readers toward fluency through increased text and concept load, less reliance on visuals, longer sentences, and more literary language.

Level 4 builds reading stamina by providing more text per page, increased use of punctuation, greater variation in sentence patterns, and increasingly challenging vocabulary.

Level 5 encourages children to move from "learning to read" to "reading to learn" by providing even more text, varied writing styles, and less familiar topics.

Whichever book is right for your reader, Blastoff! Readers are the perfect books to build confidence and encourage a love of reading that will last a lifetime!

This edition first published in 2009 by Bellwether Media, Inc.

No part of this publication may be reproduced in whole or in part without written permission of the publisher. For information regarding permission, write to Bellwether Media, Inc., Attention: Permissions Department, Post Office Box 19349, Minneapolis, MN 55419.

Library of Congress Cataloging-in-Publication Data
Wendorff, Anne.
  Lambs / by Anne Wendorff.
    p. cm. – (Blastoff! readers: watch animals grow)
  Includes bibliographical references and index.
  Summary: "A basic introduction to lambs. Developed by literacy experts with simple text and full color photography for students in kindergarten through third grade"–Provided by publisher.
  ISBN-13: 978-1-60014-241-3 (hardcover : alk. paper)
  ISBN-10: 1-60014-241-9 (hardcover : alk. paper)
  1. Lambs–Juvenile literature. I. Title.

SF376.5.W46 2009
636.3'07–dc22
                                            2008033538

Text copyright © 2009 by Bellwether Media, Inc. BLASTOFF! READERS and associated logos are trademarks and/or registered trademarks of Bellwether Media, Inc.

SCHOLASTIC, CHILDREN'S PRESS, and associated logos are trademarks and/or registered trademarks of Scholastic Inc. Printed in the United States of America.

Contents

Newborn Lambs	4
What Lambs Eat	8
Where Lambs Live	12
Lambs and Wool	16
Glossary	22
To Learn More	23
Index	24

A mother sheep has a lamb. A newborn lamb is very weak.

A lamb stands about one hour after birth. Then it learns to walk.

A newborn lamb has eight **milk teeth**. It drinks milk from its mother.

A lamb's teeth grow strong in four to five weeks. Then the lamb can **graze**.

Lambs live on
a farm.
They graze
in a field called
a **pasture**.

Lambs stay close to other sheep. The sheep move in a group called a **flock**.

Lambs are covered in hair. Their hair is called **wool**.

Lambs grow into adult sheep. They grow lots of wool.

Wool grows fast.
See how thick
it is now!

Glossary

flock—a group of sheep

graze—to eat grass and other plants that are growing in a field

milk teeth—the teeth some mammals have when they are born

pasture—a large, open area of grass and other plants

wool—thick hair covering a lamb; wool can be made into yarn or cloth.

To Learn More

AT THE LIBRARY

Greenstein, Elaine. *One Little Lamb.* New York: Viking Juvenile, 2004.

Lyon, George Ella. *Weaving the Rainbow.* New York: Atheneum/Richard Jackson Books, 2004.

Royston, Angela. *Lamb.* North Mankato, Minn.: Chrysalis Education, 2004.

ON THE WEB

Learning more about lambs is as easy as 1, 2, 3.

1. Go to www.factsurfer.com.

2. Enter "lambs" into the search box.

3. Click the "Surf" button and you will see a list of related Web sites.

With factsurfer.com, finding more information is just a click away.

Index

farm, 12
field, 12
flock, 14
graze, 10, 12
group, 14
hair, 16
hour, 6
milk, 8
milk teeth, 8
mother, 4, 8
pasture, 12
stand, 6
teeth, 10
walk, 6
weeks, 10
wool, 16, 18, 20

The images in this book are reproduced through the courtesy of: Eric Gevaert, front cover; Tom Bourdon / Alamy, p. 5; Christopher Cassidy / Alamy, p. 7; Manor Photography / Alamy, p. 9; Marilyn Barbone, p. 11; Regien Paassen, p. 13; Horizon International Images Limited / Alamy, p. 15; ChurchmouseNZ, p. 17; BESTWEB, p. 19; Ayesha Wilson, p. 21.